FOUNDATION AND SAFETY

Tasksheet Manual for NATEF Proficiency

CDX Automotive

JONES AND BARTLETT PUBLISHERS
Sudbury, Massachusetts
BOSTON TORONTO LONDON SINGAPORE

World Headquarters
Jones and Bartlett Publishers
40 Tall Pine Drive
Sudbury, MA 01776
978-443-5000
info@jbpub.com
www.jbpub.com

Jones and Bartlett Publishers
Canada
6339 Ormindale Way
Mississauga, Ontario L5V 1J2
Canada

Jones and Bartlett Publishers
International
Barb House, Barb Mews
London W6 7PA
United Kingdom

Jones and Bartlett's books and products are available through most bookstores and online booksellers. To contact Jones and Bartlett Publishers directly, call 800-832-0034, fax 978-443-8000, or visit our website, www.jbpub.com.

Substantial discounts on bulk quantities of Jones and Bartlett's publications are available to corporations, professional associations, and other qualified organizations. For details and specific discount information, contact the special sales department at Jones and Bartlett via the above contact information or send an email to specialsales@jbpub.com.

Production Credits
Chief Executive Officer: Clayton Jones
Chief Operating Officer: Don W. Jones, Jr.
President, Higher Education and Professional Publishing: Robert W. Holland, Jr.
Sr. V.P., Sales and Marketing: James Homer
V.P., Design and Production: Anne Spencer
V.P., Manufacturing and Inventory Control: Therese Connell
Publisher: Kimberly Brophy
Acquisitions Editor—Automotive: Martin Schumacher
Associate Editor: Amanda Brandt
Editorial Assistant: Kara Ebrahim
Production Manager: Jenny L. Corriveau
Director of Marketing: Alisha Weisman
Associate Marketing Manager: Jessica Cormier
Cover Design: Scott Moden
Cover Image: © Raynald Bélanger/Dreamstime.com
Composition: Shepherd, Inc.
Printing and Binding: Courier Kendallville
Cover Printing: Courier Kendallville

ISBN: 978-0-7637-8510-9

The procedures and protocols in this book are based on the most current recommendations of responsible sources. The publisher, however, makes no guarantee as to, and assume no responsibility for, the correctness, sufficiency, or completeness of such information or recommendations. Other or additional safety measures may be required under particular circumstances.

6048
Printed in the United States of America
14 13 12 11 10 9 8 7 6 5 4 3

Contents

Resource Preview

2008 CDX tasksheet numbers appear at the beginning and end of every task, as well as in the general information for the section.

Required and recommended materials and equipment are listed for each task.

Safety issues relevant to the tasks are listed at the beginning of every section.

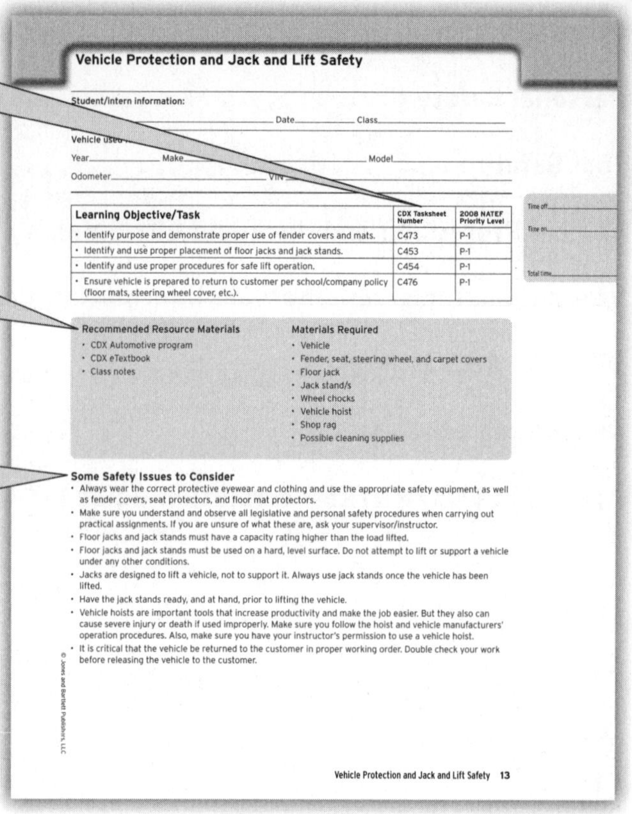

Vehicle Protection and Jack and Lift Safety

Student/intern information:

Date_____ Class_____

Vehicle used

Year_____ Make_____ Model_____

Odometer_____ VIN_____

Learning Objective/Task	CDX Tasksheet Number	2008 NATEF Priority Level
• Identify purpose and demonstrate proper use of fender covers and mats.	C473	P-1
• Identify and use proper placement of floor jacks and jack stands.	C453	P-1
• Identify and use proper procedures for safe lift operation.	C454	P-1
• Ensure vehicle is prepared to return to customer per school/company policy (floor mats, steering wheel cover, etc.).	C476	P-1

Time off_____

Time on_____

Total time_____

Recommended Resource Materials

- CDX Automotive program
- CDX eTextbook
- Class notes

Materials Required

- Vehicle
- Fender, seat, steering wheel, and carpet covers
- Floor jack
- Jack stand/s
- Wheel chocks
- Vehicle hoist
- Shop rag
- Possible cleaning supplies

Some Safety Issues to Consider

- Always wear the correct protective eyewear and clothing and use the appropriate safety equipment, as well as fender covers, seat protectors, and floor mat protectors.
- Make sure you understand and observe all legislative and personal safety procedures when carrying out practical assignments. If you are unsure of what these are, ask your supervisor/instructor.
- Floor jacks and jack stands must have a capacity rating higher than the load lifted.
- Floor jacks and jack stands must be used on a hard, level surface. Do not attempt to lift or support a vehicle under any other conditions.
- Jacks are designed to lift a vehicle, not to support it. Always use jack stands once the vehicle has been lifted.
- Have the jack stands ready, and at hand, prior to lifting the vehicle.
- Vehicle hoists are important tools that increase productivity and make the job easier. But they also can cause severe injury or death if used improperly. Make sure you follow the hoist and vehicle manufacturers' operation procedures. Also, make sure you have your instructor's permission to use a vehicle hoist.
- It is critical that the vehicle be returned to the customer in proper working order. Double check your work before releasing the vehicle to the customer.

© Jones and Bartlett Publishers, LLC

Vehicle Protection and Jack and Lift Safety 13

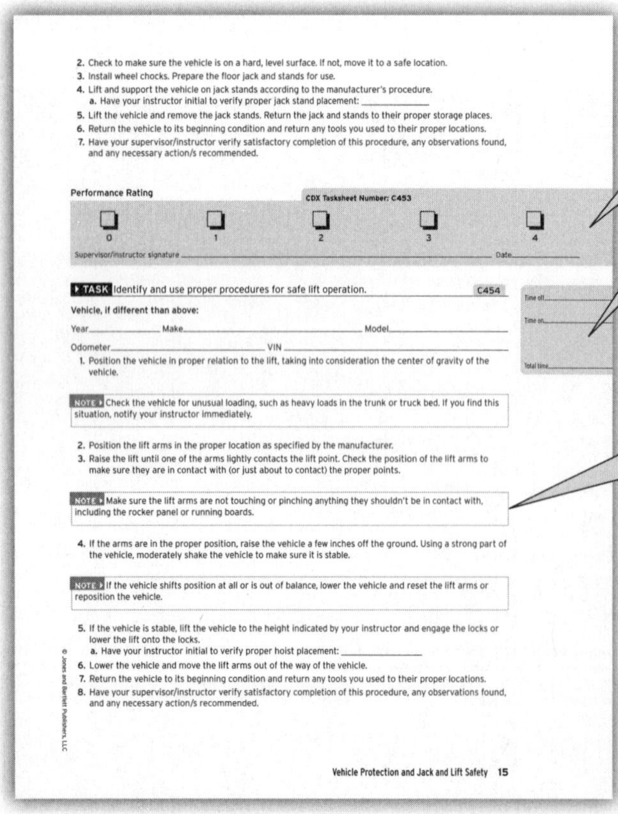

2. Check to make sure the vehicle is on a hard, level surface. If not, move it to a safe location.
3. Install wheel chocks. Prepare the floor jack and stands for use.
4. Lift and support the vehicle on jack stands according to the manufacturer's procedure.
   a. Have your instructor initial to verify proper jack stand placement: _____
5. Lift the vehicle and remove the jack stands. Return the jack and stands to their proper storage places.
6. Return the vehicle to its beginning condition and return any tools you used to their proper locations.
7. Have your supervisor/instructor verify satisfactory completion of this procedure, any observations found, and any necessary action/s recommended.

Performance Rating CDX Tasksheet Number: C453

☐ 0 ☐ 1 ☐ 2 ☐ 3 ☐ 4

Supervisor/instructor signature _____ Date_____

▸ TASK Identify and use proper procedures for safe lift operation. C454

Time off_____

Time on_____

Total time_____

Vehicle, if different than above:

Year_____ Make_____ Model_____

Odometer_____ VIN_____

1. Position the vehicle in proper relation to the lift, taking into consideration the center of gravity of the vehicle.

NOTE ▸ Check the vehicle for unusual loading, such as heavy loads in the trunk or truck bed. If you find this situation, notify your instructor immediately.

2. Position the lift arms in the proper location as specified by the manufacturer.
3. Raise the lift until one of the arms lightly contacts the lift point. Check the position of the lift arms to make sure they are in contact with (or just about to contact) the proper points.

NOTE ▸ Make sure the lift arms are not touching or pinching anything they shouldn't be in contact with, including the rocker panel or running boards.

4. If the arms are in the proper position, raise the vehicle a few inches off the ground. Using a strong part of the vehicle, moderately shake the vehicle to make sure it is stable.

NOTE ▸ If the vehicle shifts position at all or is out of balance, lower the vehicle and reset the lift arms or reposition the vehicle.

5. If the vehicle is stable, lift the vehicle to the height indicated by your instructor and engage the locks or lower the lift onto the locks.
   a. Have your instructor initial to verify proper hoist placement: _____
6. Lower the vehicle and move the lift arms out of the way of the vehicle.
7. Return the vehicle to its beginning condition and return any tools you used to their proper locations.
8. Have your supervisor/instructor verify satisfactory completion of this procedure, any observations found, and any necessary action/s recommended.

© Jones and Bartlett Publishers, LLC

Vehicle Protection and Jack and Lift Safety 15

Performance standard and rating areas simplify the sign-off process for instructors.

Time card feature appears at the beginning of every task and allows students to track the time they spend on each task.

Note boxes offer students vital information to consider while performing the task.

Acknowledgments

Jones and Bartlett Publishers would like to thank the following contributor and reviewers for their assistance with this manual:

Contributor:
Kirk VanGelder
Automotive Training Specialties
Vancouver, Washington

Reviewers:
Edward J. Heim
ASE certified in Automotive and Automotive Machinist
Battle Ground High School
Battle Ground, Washington

Marty Kamimoto
Fresno City College
Fresno, California

Sonny Reeves, CMAT, NATEF ETL
AYES Instructor
Hutchings Career Center
Macon, Georgia

Jim Stafford
Tennessee Technology Center—Newbern
Newbern, Tennessee

CDX Automotive

Designed to help automotive programs meet NATEF requirements, this interactive automotive training system introduces students to the fundamental principles and applications of automotive education. It also equips instructors with a complete set of easy-to-use course materials. This innovative resource will enrich your conventional course, or can be used as a primary management source to deliver an online program that accommodates increasing student enrollments.

CDX Automotive improves learning outcomes in automotive education by increasing student engagement and performance through personalized learning. Key features include:

- Course content mapped to the 8 ASE test areas
- Easy and intuitive navigation
- Over 850 automotive procedural and demonstration videos, including content on timely topics such as alternative fuels
- Unlimited practice tests, knowledge check exercises, and final exams that are automatically scored and recorded to save time and eliminate human error
- Support for both 2005 and 2008 NATEF standards with tasksheets and updated tasksheet reporting, making ASE recertification even easier for instructors nationwide
- *CDX Automotive Access Pack* option for students looking to purchase required or recommended access to ***CDX Automotive*** from their local college bookstore

CDX Automotive includes easy-to-use and state-of-the-art learning performance management tools to help schools manage and report student performance data that meet NATEF's ASE e-learning certification requirements. ***CDX Automotive*** offers a turnkey solution to high schools, vocational schools, community colleges and skill centers wishing to join the e-learning initiative announced by NATEF that allows programs to qualify for reduced curriculum hour requirements through e-learning completed outside of regular program hours.

Steer your course in the direction you want with ***CDX Automotive.***

Request a free trial at www.cdxauto.com.

General Safety

Student/intern information:

Name________________ Date________ Class________________

Time off________

Time on________

Total time________

Learning Objective/Task	CDX Tasksheet Number	2008 NATEF Priority Level
• Identify general shop rules and procedures.	C451	P-1
• Identify marked safety areas.	C456	P-1
• Identify the location and use of fire blankets.	C457	P-1
• Identify the types and location of fire extinguishers; demonstrate knowledge of the procedures for using fire extinguishers.	C458	P-1
• Identify the location and use of eye wash stations.	C459	P-1
• Identify the location of posted evacuation routes.	C460	P-1
• Locate and demonstrate knowledge of material safety data sheets (MSDS).	C465	P-1

Recommended Resource Materials

- CDX Automotive program
- CDX *e*Textbook
- Class notes

Materials Required

- Program's shop policy and other safety information
- MSDS book
- Fire extinguisher(s) from the shop
- Material Safety Data Sheets for the products listed

Some Safety Issues to Consider

- Always wear the correct protective eyewear and clothing and use the appropriate safety equipment, as well as fender covers, seat protectors, and floor mat protectors.
- Make sure you understand and observe all legislative and personal safety procedures when carrying out practical assignments. If you are unsure of what these are, ask your supervisor/instructor.
- Shop rules and procedures are critical to your safety. Please give this your utmost attention.
- Marked safety areas play an important role in maintaining a safe work environment. Always understand and heed marked safety areas.
- Fire blankets are an important component of fire safety in the shop. Always know their location, purpose, and use.
- Fire extinguishers come in a variety of types and sizes. It pays to understand the differences before they are needed.
- Eye wash stations are an important component of shop safety. Always know their location, purpose, and use.
- Pre-planned evacuation routes are an important component of shop safety. Always know the location of all evacuation routes for your shop.
- Being able to locate information in an MSDS is critical to your safety and health. Make sure you are familiar with their location and use.

Performance Standard

0–No exposure: No information or practice provided during the program; complete training required

1–Exposure only: General information provided with no practice time; close supervision needed; additional training required

2–Limited practice: Has practiced job during training program; additional training required to develop skill

3–Moderately skilled: Has performed job independently during training program; limited additional training may be required

4–Skilled: Can perform job independently with no additional training

Total time________

Time on________

Time off________

▶ TASK Identify general shop rules and procedures. C451

1. List the location/s of the following items.
   a. Program's general shop rules and policies: ________
   b. Material Data Safety Sheets (MSDS) book: ________
   c. Procedure for operation of a fire extinguisher: ________
   d. Procedure for operation of a vehicle hoist: ________
2. List the shop's policy for wearing of safety glasses while in the shop:
3. List the shop's policy for driving vehicles:
4. List the shop's policy for clothing in the shop:
5. List the shop's policy for jewelry in the shop:
6. Pass the shop's safety test and record your score here: ________
7. Have your supervisor/instructor verify satisfactory completion of this procedure, any observations found, and any necessary action/s recommended.

Performance Rating

CDX Tasksheet Number: C451

☐	☐	☐	☐	☐
0	1	2	3	4

Supervisor/instructor signature ________ Date________

Total time________

Time on________

Time off________

▶ TASK Identify marked safety areas. C456

1. Research the uniform color code system used to designate safety areas in a shop.
   List each color and its designation:
   a. Color: ________ designates: ________
   b. Color: ________ designates: ________
   c. Color: ________ designates: ________
   d. Color: ________ designates: ________
2. Label all of the marked safety areas on the diagram of the shop at the end of this section.
3. Have your supervisor/instructor verify satisfactory completion of this procedure, any observations found, and any necessary action/s recommended.

Performance Rating

CDX Tasksheet Number: C456

☐ 0 ☐ 1 ☐ 2 ☐ 3 ☐ 4

Supervisor/instructor signature _______________ Date_______

▶ TASK Identify the location and use of fire blankets. **C457**

Time off_______

Time on_______

Total time_______

1. Research the location, purpose, and use of fire blankets in your shop.
   a. Describe the purpose of a fire blanket (when should a fire blanket be used?):

   b. Describe how you should use a fire blanket:

2. Label the location of the fire blanket(s) on the diagram of the shop at the end of this section.
3. Have your supervisor/instructor verify satisfactory completion of this procedure, any observations found, and any necessary action/s recommended.

Performance Rating

CDX Tasksheet Number: C457

☐ 0 ☐ 1 ☐ 2 ☐ 3 ☐ 4

Supervisor/instructor signature _______________ Date_______

▶ TASK Identify the types and location of fire extinguishers; demonstrate knowledge of the procedures for using fire extinguishers. **C458**

Time off_______

Time on_______

Total time_______

1. Research the types, location, and use of fire extinguishers.
   a. List the different types of fire extinguishers (dry chemical, CO_2, etc.) available:
      i. Type: _______________; for use on what class of fire: _______________
      ii. Type: _______________; for use on what class of fire: _______________
      iii. Type: _______________; for use on what class of fire: _______________
      iv. Type: _______________; for use on what class of fire: _______________
      v. In the above list, put a star next to the type(s) of fire extinguishers in this shop.
2. List the steps for proper use of a fire extinguisher:

3. Label the location of all fire extinguishers on the diagram of the shop at the end of this section.
4. Have your supervisor/instructor verify satisfactory completion of this procedure, any observations found, and any necessary action/s recommended.

Performance Rating

CDX Tasksheet Number: C458

☐ 0 ☐ 1 ☐ 2 ☐ 3 ☐ 4

Supervisor/instructor signature ______________________ Date______

Total time______

Time on______

Time off______

▶ TASK Identify the location and use of eye wash stations. **C459**

1. Research the location, purpose, and use of eye wash stations in your shop.
   a. Describe the purpose of an eye wash station:

   b. Describe the proper use of an eye wash station (including time):

   c. What type of eye injury would use of an eye wash station not be appropriate for?

2. Label the location of the eye wash station/s on the diagram of the shop at the end of this section.
3. Have your supervisor/instructor verify satisfactory completion of this procedure, any observations found, and any necessary action/s recommended.

Performance Rating

CDX Tasksheet Number: C459

☐ 0 ☐ 1 ☐ 2 ☐ 3 ☐ 4

Supervisor/instructor signature ______________________ Date______

Total time______

Time on______

Time off______

▶ TASK Identify the location of posted evacuation routes. **C460**

1. Research the location and purpose of evacuation routes.
   a. Describe the purpose of evacuation routes:

2. Label the evacuation route/s on the diagram of the shop at the end of this section. Also label the location of the posted evacuation route diagram/s.
3. Have your supervisor/instructor verify satisfactory completion of this procedure, any observations found, and any necessary action/s recommended.

Performance Rating

CDX Tasksheet Number: C460

☐ 0 ☐ 1 ☐ 2 ☐ 3 ☐ 4

Supervisor/instructor signature ______________________________ Date______________

▶ TASK Locate and demonstrate knowledge of material safety data sheets (MSDS). **C465**

Time off______________

Time on______________

Total time______________

1. List the safety precautions when handling motor oil:

2. List the flash point of gasoline/petroleum: __________________ °F/°C
3. List the firefighting equipment needed to put out a gasoline/petroleum fire:

4. List the first aid treatment for battery acid in the eyes:

5. List the first aid treatment for ingestion of anti-freeze (ethylene glycol):

6. Label the location of the MSDS books on the diagram of the shop at the end of this section.
7. Have your supervisor/instructor verify satisfactory completion of this procedure, any observations found, and any necessary action/s recommended.

Performance Rating

CDX Tasksheet Number: C465

☐ 0 ☐ 1 ☐ 2 ☐ 3 ☐ 4

Supervisor/instructor signature ______________________________ Date______________

Draw a diagram of the shop and label the following:

- Marked safety areas (floor lines, etc.)
- Location of the fire blanket/s
- Location of all fire extinguishers
- Location of the exhaust fan switch/es
- Location of the electrical disconnect switch/es
- Location of the eye wash station/s
- Location of the MSDS books
- Evacuation route/s
- Location of the master air shut-off valve
- Location of the over-head door switch/es

Personal Safety

Student/intern information:

Name________________ Date________ Class________________

Learning Objective/Task	CDX Tasksheet Number	2008 NATEF Priority Level
• Comply with the safe use of safety glasses, gloves, and shoes during lab/shop activities.	C461	P-1
• Identify and wear appropriate clothing for lab/shop activities.	C462	P-1
• Comply with appropriate hairstyles for lab/shop activities.	C463	P-1

Time off________

Time on________

Total time________

Recommended Resource Materials

- CDX General Service
- CDX Automotive program
- CDX *e*Textbook
- Class notes

Materials Required

- Regular safety glasses
- Gloves
- Shoes

Some Safety Issues to Consider

- Make sure you understand and observe all legislative and personal safety procedures when carrying out practical assignments. If you are unsure of what these are, ask your supervisor/instructor.

Performance Standard

0–No exposure: No information or practice provided during the program; complete training required
1–Exposure only: General information provided with no practice time; close supervision needed; additional training required
2–Limited practice: Has practiced job during training program; additional training required to develop skill
3–Moderately skilled: Has performed job independently during training program; limited additional training may be required
4–Skilled: Can perform job independently with no additional training

▸ TASK Comply with the safe use of safety glasses, gloves, and shoes during lab/shop activities. **C461**

▸ TASK Identify and wear appropriate clothing for lab/shop activities. **C462**

▸ TASK Comply with appropriate hairstyles for lab/shop activities. **C463**

Time off________

Time on________

Total time________

1. Describe the safety glasses policy for your shop (be specific):

2. Describe the policy related to gloves for your shop (be specific):

3. Describe the policy related to work shoes for your shop (be specific):

4. Describe the clothing requirements for your shop (be specific):

5. List your shop's policy concerning hair safety in the shop (be specific):

6. These tasks require observation of the student over a prolonged period. Ask your instructor to give you a date for your evaluation.
   a. Write that date here: ______________________
7. Continue with your projects, complying with the safe use of safety glasses, gloves, shoes, clothing, and hair containment during all lab/shop activities.
8. On or after that date, have your instructor verify satisfactory completion of these tasks.

Performance Rating | **CDX Tasksheet Number: C461**

☐ 0 ☐ 1 ☐ 2 ☐ 3 ☐ 4

Supervisor/instructor signature ______________________ Date______

Performance Rating | **CDX Tasksheet Number: C462**

☐ 0 ☐ 1 ☐ 2 ☐ 3 ☐ 4

Supervisor/instructor signature ______________________ Date______

Performance Rating | **CDX Tasksheet Number: C463**

☐ 0 ☐ 1 ☐ 2 ☐ 3 ☐ 4

Supervisor/instructor signature ______________________ Date______

Tool Safety

Student/intern information:

Name____________________ Date__________ Class________________

Learning Objective/Task	CDX Tasksheet Number	2008 NATEF Priority Level
• Identify tools and their usage in automotive applications.	C466/C504	P-1
• Identify standard and metric designation.	C467	P-1
• Demonstrate safe handling and use of appropriate tools.	C468	P-1
• Utilize safe procedures for handling tools and equipment.	C452	P-1
• Demonstrate proper cleaning, storage, and maintenance of tools and equipment.	C469	P-1

Time off________

Time on________

Total time________

Recommended Resource Materials
- CDX General Service
- CDX Automotive program
- CDX *e*Textbook
- Class notes

Materials Required
- Standard toolkit and other tools as required

Some Safety Issues to Consider
- Always wear the correct protective eyewear and clothing and use the appropriate safety equipment, as well as fender covers, seat protectors, and floor mat protectors.
- Make sure you understand and observe all legislative and personal safety procedures when carrying out practical assignments. If you are unsure of what these are, ask your supervisor/instructor.
- Tools allow us to increase our productivity and effectiveness. However, they must be used according to the manufacturer's procedures. Failure to follow those procedures can result in serious injury or death.

Performance Standard
0–No exposure: No information or practice provided during the program; complete training required
1–Exposure only: General information provided with no practice time; close supervision needed; additional training required
2–Limited practice: Has practiced job during training program; additional training required to develop skill
3–Moderately skilled: Has performed job independently during training program; limited additional training may be required
4–Skilled: Can perform job independently with no additional training

▶ TASK Identify tools and their usage in automotive applications. **C466/C504**

Time off________

Time on________

Total time________

▶ TASK Identify standard and metric designation. **C467**

▶ TASK Demonstrate safe handling and use of appropriate tools. **C468**

▶ TASK Utilize safe procedures for handling tools and equipment. **C452**

▶ TASK Demonstrate proper cleaning, storage, and maintenance of tools and equipment. **C469**

1. This task will require observation of the student over a prolonged period after the initial check. Ask your instructor to give you a date for your evaluation.
   a. Write that date here: ________________

2. Using the following list, describe the specific function/purpose of each of the following tools:
   a. Open end wrench: ______
   b. Box end wrench: ______
   c. Socket: ______
   d. Ratchet: ______
   e. Torque wrench: ______
   f. Slotted screwdriver: ______
   g. Phillips screwdriver: ______
   h. Tap: ______
   i. Die: ______
   j. Feeler blade: ______
   k. Line wrench/flare nut wrench: ______
   l. Allen wrench: ______
   m. Torx screwdriver or socket: ______
   n. Hacksaw: ______
   o. Oil filter wrench: ______
   p. Compression gauge: ______
   q. DVOM/DMM: ______
   r. Test light: ______
   s. Diagonal side cutters: ______
   t. Locking pliers: ______
   u. Needle-nose pliers: ______
   v. Brake spoon: ______
   w. Micrometer: ______
   x. Dial indicator: ______
   y. Antifreeze hydrometer: ______
   z. Snap ring pliers: ______
3. Continue with your other projects, demonstrating safe handling, proper cleaning, maintenance, and storage of the tools until the date of your evaluation.
4. On or after that date, have your instructor verify satisfactory completion of this task.
5. Using the following tables, continue with your other projects, demonstrating your understanding of the differences in the designations until the date of your final evaluation.
6. On or after that date, have your instructor verify satisfactory completion of this task.

Volume

Volume is the amount of space occupied by a three-dimensional object. The metric system uses liters (l) or cubic centimeters (cc or cm^3). The Imperial system uses gallons and quarts for "wet" volume and cubic feet for "dry" volume. You'll need to determine volume any time you fill a vehicle's reservoir with liquid. This includes petrol/gasoline, coolant, oil, transmission fluid, or lubricant.

Volume Conversions	
Imperial-Imperial	4 US quarts = 1 US gallon 1 cubic foot = 7.48 US gallons 1 cubic foot = 6.22 UK gallons
Metric-Metric	1 liter = 1000 cc 1 cc = 0.001 liter
Imperial-Metric	1 cubic inch = 16.387 cc 1 US gallon = 3.78 liters 1 UK gallon = 4.54 liters 1 US quart = 0.95 liters
Metric-Imperial	1 liter = 61.0237 cubic inches 1 liter = 0.035 cubic feet 1 liter = 0.26 US gallons 1 liter = 0.21 UK gallons 1 liter = 1.05 quarts

7. Knowledge Check: Convert the following:
   a. 3.0 liters = __________ cubic inches
   b. 350 cubic inches = ______________ liters
   c. 3.0 gallons = _________ liters
   d. 9 liters = _____________ gallons

Mass

Mass is a unit or system of units by which a degree of heaviness is measured. The metric system uses grams (g), kilograms (kg), and tonnes (t). The Imperial system uses ounces (oz), pounds (lb), and tons (T). In the workshop, you will use these measurements to determine the lifting capacity of equipment like hydraulic and engine hoists and floor jacks.

Mass Conversions	
Imperial-Imperial	16 oz = 1 lb 2000 lb = 1 T
Metric-Metric	1000 g = 1 kg 1000 kg = 1 t
Imperial-Metric	1 oz = 28.3 g 1 lb = 453 g 2.2 lb = 1 kg 1 T = 0.907 t
Metric-Imperial	1 t = 1.10 T

8. Knowledge Check: Convert the following:
   a. 8 oz = ___________ g
   b. 475 g = _____________ oz
   c. 6.6 lb = ____________ kg
   d. 4500 kg = _____________ lb

Torque

Torque is the twisting force applied to a shaft. The metric system uses the Newton meter (Nm). The Imperial system uses the inch-pound (in-lb) and the foot-pound (ft-lb). Vehicle manufacturers specify torque settings for key fasteners on the engine and wheels. You will need to follow the specifications or you could strip threads or break bolts. Torque is also an important concept when discussing engine performance. A foot-pound (ft-lb) is the twisting force applied to a shaft by a lever 1 foot long with a 1 pound mass on the end. A Newton meter (Nm) is the twisting force applied to a shaft by a level 1 meter long with a force of 1 Newton applied to the end of the lever. (1N is equivalent to the force applied by a mass of 100.)

Torque Conversions	
Imperial-Imperial	12 in-lb = 1 ft-lb 1 in-lb = 0.08 ft-lb
Imperial-Metric	1 ft-lb = 1.34 Nm
Metric-Imperial	1 Nm = 0.74 ft-lb 1 Nm = 8.8 in-lb

9. Knowledge Check: Convert the following:
   a. 48 in-lb = ________ ft-lb
   b. 15 ft-lb = _____________ in-lb
   c. 65 ft-lb = _________ Nm
   d. 142 Nm = _____________ ft-lb

Pressure

Pressure is a measurement of force per unit area. The metric system uses kilopascals (kPa) and bar. The Imperial system uses pounds per square inch (psi) and atmospheres. *Vacuum* is a term given to a pressure that is less than atmospheric pressure. The Imperial system measures vacuum in inches of mercury ("Hg) or inches of water. The metric system measures vacuum in millimeters of mercury (mm Hg). You'll need to understand pressure conversions when filling tires and replacing air conditioning refrigerants or using a vacuum gauge.

Pressure Conversions	
Imperial-Imperial	14.7 psi = 1 atmosphere 1" Hg = 14"H_2O 0" Hg = 1 atmosphere
Metric-Metric	100 kPa = 1 bar
Imperial-Metric	1 psi = 6.89 kPa 1 atmosphere = 101.3 kPa 1" Hg = 25.4 mm Hg 1 atmosphere = 1.013 bar

10. Knowledge Check: Convert the following:
    a. 14.7 psi = __________ kPa
    b. 650 kPa = ____________ psi
    c. 22 psi = __________ bar
    d. 5.5 bar = ____________ psi
11. Have your supervisor/instructor verify satisfactory completion of this procedure, any observations found, and any necessary action/s recommended.

Performance Rating

CDX Tasksheet Number: C466/504, 467, 468, 452, and 469

☐ 0 ☐ 1 ☐ 2 ☐ 3 ☐ 4

Supervisor/instructor signature ______________________ Date ________

Vehicle Protection and Jack and Lift Safety

Student/intern information:

Name________________ Date________ Class____________

Vehicle used for this activity:

Year________ Make________________ Model____________

Odometer________________ VIN ____________________

Time off________

Time on________

Total time________

Learning Objective/Task	CDX Tasksheet Number	2008 NATEF Priority Level
• Identify purpose and demonstrate proper use of fender covers and mats.	C473	P-1
• Identify and use proper placement of floor jacks and jack stands.	C453	P-1
• Identify and use proper procedures for safe lift operation.	C454	P-1
• Ensure vehicle is prepared to return to customer per school/company policy (floor mats, steering wheel cover, etc.).	C476	P-1

Recommended Resource Materials

- CDX Automotive program
- CDX eTextbook
- Class notes

Materials Required

- Vehicle
- Fender, seat, steering wheel, and carpet covers
- Floor jack
- Jack stand/s
- Wheel chocks
- Vehicle hoist
- Shop rag
- Possible cleaning supplies

Some Safety Issues to Consider

- Always wear the correct protective eyewear and clothing and use the appropriate safety equipment, as well as fender covers, seat protectors, and floor mat protectors.
- Make sure you understand and observe all legislative and personal safety procedures when carrying out practical assignments. If you are unsure of what these are, ask your supervisor/instructor.
- Floor jacks and jack stands must have a capacity rating higher than the load lifted.
- Floor jacks and jack stands must be used on a hard, level surface. Do not attempt to lift or support a vehicle under any other conditions.
- Jacks are designed to lift a vehicle, not to support it. Always use jack stands once the vehicle has been lifted.
- Have the jack stands ready, and at hand, prior to lifting the vehicle.
- Vehicle hoists are important tools that increase productivity and make the job easier. But they also can cause severe injury or death if used improperly. Make sure you follow the hoist and vehicle manufacturers' operation procedures. Also, make sure you have your instructor's permission to use a vehicle hoist.
- It is critical that the vehicle be returned to the customer in proper working order. Double check your work before releasing the vehicle to the customer.

Performance Standard

0–No exposure: No information or practice provided during the program; complete training required

1–Exposure only: General information provided with no practice time; close supervision needed; additional training required

2–Limited practice: Has practiced job during training program; additional training required to develop skill

3–Moderately skilled: Has performed job independently during training program; limited additional training may be required

4–Skilled: Can perform job independently with no additional training

Total time________

Time on________

Time off________

▶ TASK Identify purpose and demonstrate proper use of fender covers and mats. **C473**

1. Identify the purpose of the following items:
   a. Fender cover:

   b. Seat cover:

   c. Steering wheel cover:

   d. Carpet cover:

2. Properly prepare a vehicle for service or repair, using the above covers.
3. Have your supervisor/instructor verify satisfactory completion of this procedure, any observations found, and any necessary action/s recommended.

Performance Rating

CDX Tasksheet Number: C473

☐ 0 ☐ 1 ☐ 2 ☐ 3 ☐ 4

Supervisor/instructor signature ________________ Date________

Total time________

Time on________

Time off________

▶ TASK Identify and use proper placement of floor jacks and jack stands. **C453**

1. Research the jacking and lifting procedures for this vehicle in the appropriate service manual.
   a. Draw a diagram of the vehicle's lift points:

2. Check to make sure the vehicle is on a hard, level surface. If not, move it to a safe location.
3. Install wheel chocks. Prepare the floor jack and stands for use.
4. Lift and support the vehicle on jack stands according to the manufacturer's procedure.
   a. Have your instructor initial to verify proper jack stand placement: ________________
5. Lift the vehicle and remove the jack stands. Return the jack and stands to their proper storage places.
6. Return the vehicle to its beginning condition and return any tools you used to their proper locations.
7. Have your supervisor/instructor verify satisfactory completion of this procedure, any observations found, and any necessary action/s recommended.

Performance Rating

CDX Tasksheet Number: C453

☐	☐	☐	☐	☐
0	1	2	3	4

Supervisor/instructor signature ________________________________ Date________________

▶ TASK Identify and use proper procedures for safe lift operation. **C454**

Time off________________

Time on________________

Total time________________

Vehicle, if different than above:

Year______________ Make________________________________ Model____________________

Odometer______________________________ VIN ______________________________

1. Position the vehicle in proper relation to the lift, taking into consideration the center of gravity of the vehicle.

> **NOTE ▶** Check the vehicle for unusual loading, such as heavy loads in the trunk or truck bed. If you find this situation, notify your instructor immediately.

2. Position the lift arms in the proper location as specified by the manufacturer.
3. Raise the lift until one of the arms lightly contacts the lift point. Check the position of the lift arms to make sure they are in contact with (or just about to contact) the proper points.

> **NOTE ▶** Make sure the lift arms are not touching or pinching anything they shouldn't be in contact with, including the rocker panel or running boards.

4. If the arms are in the proper position, raise the vehicle a few inches off the ground. Using a strong part of the vehicle, moderately shake the vehicle to make sure it is stable.

> **NOTE ▶** If the vehicle shifts position at all or is out of balance, lower the vehicle and reset the lift arms or reposition the vehicle.

5. If the vehicle is stable, lift the vehicle to the height indicated by your instructor and engage the locks or lower the lift onto the locks.
   a. Have your instructor initial to verify proper hoist placement: ________________
6. Lower the vehicle and move the lift arms out of the way of the vehicle.
7. Return the vehicle to its beginning condition and return any tools you used to their proper locations.
8. Have your supervisor/instructor verify satisfactory completion of this procedure, any observations found, and any necessary action/s recommended.

Performance Rating

CDX Tasksheet Number: C454

☐ 0 ☐ 1 ☐ 2 ☐ 3 ☐ 4

Supervisor/instructor signature ______________________________ Date______________

Total time______________

Time on______________

Time off______________

▶ TASK Ensure vehicle is prepared to return to customer per school/company policy (floor mats, steering wheel cover, etc.). **C476**

NOTE ▶ A properly protected vehicle and good work habits will make this task much easier.

1. Double check that all work has been completed. Nothing can be missing or loose.
   a. Student initial when completed: ______________
2. Double check that all tools are put away and stored properly.
   a. Student initial when completed: ______________
3. If your instructor deems it necessary, test drive the vehicle to be sure of proper repair and operation of the vehicle.
   a. Have your instructor initial here: ______________
4. If the vehicle is ready to return to the customer, the vehicle may need to be moved out of the shop. Get your instructor's permission to move the vehicle to the customer pick-up area.
   a. Have your instructor initial here: ______________
5. Remove all fender covers, seat covers, floor covers, and steering wheel covers. Return them to their storage place or dispose of them, depending on the type of cover.
   a. Student initial when completed: ______________
6. Check the exterior of the vehicle for greasy fingerprints or grime. Clean with an appropriate cleaner. Follow your shop's policies on this procedure.
   a. Student initial when completed: ______________
7. Check the following interior locations for dirt or greasy spots. Clean with an appropriate cleaner. Follow your shop's policies on this procedure:
   a. Carpet and floor mats. Student initial when completed: ______________
   b. Seats. Student initial when completed: ______________
   c. Steering wheel and parking brake handle. Student initial when completed: __________
   d. Door panel and handles. Student initial when completed: ______________
8. Return to your work stall and clean up the floor, benches, and related area.
   a. Student initial when completed: ______________
9. Have your supervisor/instructor verify satisfactory completion of this procedure, any observations found, and any necessary action/s recommended.

Performance Rating

CDX Tasksheet Number: C476

☐ 0 ☐ 1 ☐ 2 ☐ 3 ☐ 4

Supervisor/instructor signature ______________________________ Date______________

Vehicle, Customer, and Service Information

Student/intern information:

Name____________________ Date__________ Class____________

Vehicle used for this activity:

Year__________ Make____________________ Model____________

Odometer____________________ VIN ____________________

Learning Objective/Task	CDX Tasksheet Number	2008 NATEF Priority Level
• Proper vehicle identification information: Define the use and purpose of the VIN, engine numbers and data code, locate VIN, apply knowledge of VIN information.	C471	P-1
• Identify information needed and the service requested on a repair order.	C472	P-1
• Demonstrate use of 3 Cs (concern, cause, and correction).	C474	P-1
• Review vehicle service history.	C475	P-1

Time off__________

Time on__________

Total time__________

Recommended Resource Materials

- CDX General Service
- CDX Automotive program
- CDX *e*Textbook
- Technical service bulletins, shop manuals, and any other information applicable to the specific vehicle or components you are working on
- Class notes

Materials Required

- Vehicle
- Service information
- Completed repair order assigned by your instructor
- Several repair orders for the same vehicle from repairs/services performed over an extended period of time
- Scheduled maintenance chart for this vehicle

Some Safety Issues to Consider

- Always wear the correct protective eyewear and clothing and use the appropriate safety equipment, as well as fender covers, seat protectors, and floor mat protectors.
- Make sure you understand and observe all legislative and personal safety procedures when carrying out practical assignments. If you are unsure of what these are, ask your supervisor/instructor.

Performance Standard

0–No exposure: No information or practice provided during the program; complete training required

1–Exposure only: General information provided with no practice time; close supervision needed; additional training required

2–Limited practice: Has practiced job during training program; additional training required to develop skill

3–Moderately skilled: Has performed job independently during training program; limited additional training may be required

4–Skilled: Can perform job independently with no additional training

Total time_______
Time on_______
Time off_______

▶ TASK Proper vehicle identification information: Define the use and purpose of the VIN, engine numbers and data code, locate VIN, apply knowledge of VIN information. **C471**

1. Research the location and description of the VIN in an appropriate service manual.
   a. Location of VIN: ________________________
2. Using the VIN on the vehicle assigned to you, identify the following information.
   a. Country of origin: ________________________
   b. Location of plant: ________________________
   c. Passenger restraint system: ________________________
   d. Engine designation: ________________________
   e. Model year: ______________
3. Have your supervisor/instructor verify satisfactory completion of this procedure, any observations found, and any necessary action/s recommended.

Performance Rating

CDX Tasksheet Number: C471

☐ 0 ☐ 1 ☐ 2 ☐ 3 ☐ 4

Supervisor/instructor signature ________________________ Date________

Total time_______
Time on_______
Time off_______

▶ TASK Identify information needed and the service requested on a repair order. **C472**

1. Familiarize yourself with the assigned repair order. Locate and list the following information below.
   a. Date: ______________
   b. Customer: ________________________
   c. Address: ________________________
   d. Daytime phone number: ________________________
   e. Year: __________
   f. Make: ________________________
   g. Model: ________________________
   h. Color: ________________
   i. Odometer reading: ________________
   j. VIN: ________________________
   k. Customer concern(s)/service requested:
2. Did the customer sign the repair order authorizing the repairs? **Yes/No** (Circle one)
3. Return the sample repair order to its proper storage place.
4. Have your supervisor/instructor verify satisfactory completion of this procedure, any observations found, and any necessary action/s recommended.

Performance Rating

CDX Tasksheet Number: C472

☐ 0 ☐ 1 ☐ 2 ☐ 3 ☐ 4

Supervisor/instructor signature ________________________ Date________

▸ TASK Demonstrate use of 3 Cs (concern, cause, and correction). C474

Time off________

Time on________

Total time________

1. Using the following scenario, write up the 3 Cs as listed on most repair orders. Assume that the customer authorized the recommended repairs.

   A vehicle is towed into your shop with a no-start condition. The engine cranks over very slowly. You test the battery and find it is fully charged. You test the starter and find that it is drawing excessive battery current, but sounds normal other than the cranking speed. You check the oil level and find it to be 3 quarts low (4 quart capacity). You turn the crankshaft by hand and find it extremely hard to turn. The customer authorizes the removal of the oil pan to inspect the crankshaft and bearings. You find a spun #2 main bearing and the crankshaft heavily scored, requiring replacement. You also notice severe piston and cylinder wall damage.

2. Concern/complaint:

3. Cause:

4. Correction:

5. Have your supervisor/instructor verify satisfactory completion of this procedure, any observations found, and any necessary action/s recommended.

Performance Rating

CDX Tasksheet Number: C474

☐	☐	☐	☐	☐
0	1	2	3	4

Supervisor/instructor signature ________________ Date________

▸ TASK Review vehicle service history. C475

Time off________

Time on________

Total time________

1. Familiarize yourself with the repair history as listed on the repair orders and answer the following questions.
   a. When was the first date this vehicle was serviced? ________________
   b. What was the last date this vehicle was serviced? ________________
   c. What was the most major repair performed? ________________
   d. Was this vehicle ever returned for the same problem more than once? **Yes/No** (Circle one)
      i. If so, for what and how many times? ________________
   e. Compare this list to the scheduled maintenance chart and list any missed maintenance tasks between the first service and the last service:

2. Have your supervisor/instructor verify satisfactory completion of this procedure, any observations found, and any necessary action/s recommended.

Performance Rating

CDX Tasksheet Number: C475

☐	☐	☐	☐	☐
0	1	2	3	4

Supervisor/instructor signature ________________________________ Date________

SRS and ABS Safety

Student/intern information:

Name______________________________ Date____________ Class____________________

Vehicle used for this activity:

Year____________ Make______________________________ Model____________________

Odometer____________________________ VIN ______________________________

Learning Objective/Task	**CDX Tasksheet Number**	**2008 NATEF Priority Level**
• Demonstrate knowledge of safety aspects of supplemental restraint systems (SRS) and anti-lock braking systems (ABS).	C464	P-1

Time off____________

Time on____________

Total time____________

Recommended Resource Materials

- CDX Automotive program
- CDX eTextbook
- Technical service bulletins, shop manuals, and any other information applicable to the specific vehicle or components you are working on
- Class notes

Materials Required

- Vehicle with SRS and ABS
- Special tools or equipment, if specified

Some Safety Issues to Consider

- Always wear the correct protective eyewear and clothing and use the appropriate safety equipment, as well as fender covers, seat protectors, and floor mat protectors.
- Make sure you understand and observe all legislative and personal safety procedures when carrying out practical assignments. If you are unsure of what these are, ask your supervisor/instructor.
- Deployment of the SRS system could cause serious injury or death. Follow all of the manufacturer's procedures before working on this system.
- The ABS system may hold brake fluid under extremely high pressure. Follow the manufacturer's procedures to relieve this pressure before working on this system.

Performance Standard

0–No exposure: No information or practice provided during the program; complete training required
1–Exposure only: General information provided with no practice time; close supervision needed; additional training required
2–Limited practice: Has practiced job during training program; additional training required to develop skill
3–Moderately skilled: Has performed job independently during training program; limited additional training may be required
4–Skilled: Can perform job independently with no additional training

Total time________

Time on________

Time off________

TASK Demonstrate knowledge of safety aspects of supplemental restraint systems (SRS) and anti-lock braking systems (ABS). **C464**

1. Research the following procedures for this vehicle in the appropriate service manual.
   a. List the steps to disable the SRS system on this vehicle:

   b. List the steps to relieve the pressure in the ABS accumulator, if applicable:

2. Following the manufacturer's procedure, disable the SRS system.
3. Following the manufacturer's procedure, relieve the ABS pressure.
   a. Have your instructor initial to verify the SRS is disabled and ABS pressure is relieved:

      ____
4. Return the SRS and ABS systems to proper operating status.
   a. Have your supervisor/instructor initial to verify proper operating status of SRS and ABS: ________________
5. Return the vehicle to its beginning condition and return any special tools or equipment that you may have used to their proper locations.
6. Have your instructor/supervisor verify satisfactory completion of this procedure, any observations found, and any necessary action/s recommended.

Performance Rating

CDX Tasksheet Number: C464

☐	☐	☐	☐	☐
0	1	2	3	4

Supervisor/instructor signature ________________________________ Date________________

Associated Studies

Student/intern information:

Name________________ Date________ Class________

Time off________

Time on________

Total time________

Learning Objective/Task	CDX Tasksheet Number	2008 NATEF Priority Level
• Recognize different vehicle types and body components.	C505	P-1

Recommended Resource Materials

- CDX Automotive program
- CDX *e*Textbook
- Technical service bulletins, shop manuals, and any other information applicable to the specific vehicle or components you are working on
- Class notes

Materials Required

- A range of vehicles of the designated types which are fitted with different configurations

Some Safety Issues to Consider

- Always wear the correct protective eyewear and clothing and use the appropriate safety equipment, as well as fender covers, seat protectors, and floor mat protectors.
- Make sure you understand and observe all legislative and personal safety procedures when carrying out practical assignments. If you are unsure of what these are, ask your supervisor/instructor.

Performance Standard

0–No exposure: No information or practice provided during the program; complete training required

1–Exposure only: General information provided with no practice time; close supervision needed; additional training required

2–Limited practice: Has practiced job during training program; additional training required to develop skill

3–Moderately skilled: Has performed job independently during training program; limited additional training may be required

4–Skilled: Can perform job independently with no additional training

▶ TASK Recognize different vehicle types and body components. C505

Time off________

Time on________

Total time________

1. Inspect a range of vehicles or illustrations (including computer-based), identify the following types of vehicles, and complete the table below.

Vehicle Type	Year	Make	Model
Convertible			
Four-wheel drive			
Cube van			
Minivan			
Front-engine RWD			
Mid-engine RWD			
Rear-engine RWD			
Front-engine FWD			

2. Now inspect a range of vehicles or illustrations (including computer-based), identify the following vehicle body parts, and complete the table below.

Component	Year	Make	Model
Sun roof			
Tailgate			
Rear spoiler			
Hinged doors			
Sliding doors			
Satellite navigation			
Aerodynamic headlights			

3. Have your supervisor/instructor verify satisfactory completion of this procedure, any observations found, and any necessary action/s recommended.

Performance Rating

CDX Tasksheet Number: C505

☐ 0 ☐ 1 ☐ 2 ☐ 3 ☐ 4

Supervisor/instructor signature ______________________ Date______

Student/intern information:

Name______________________________ Date__________ Class____________________

Learning Objective/Task	CDX Tasksheet Number	2008 NATEF Priority Level
• Complete an accident report.	C593	P-1

Time off__________

Time on__________

Total time__________

Recommended Resource Materials
- CDX General Service
- CDX Automotive program
- CDX eTextbook
- Class notes

Materials Required
- School accident report

Some Safety Issues to Consider

- Make sure you understand and observe all legislative and personal safety procedures when carrying out practical assignments. If you are unsure of what these are, ask your supervisor/instructor.

Performance Standard

0–No exposure: No information or practice provided during the program; complete training required
1–Exposure only: General information provided with no practice time; close supervision needed; additional training required
2–Limited practice: Has practiced job during training program; additional training required to develop skill
3–Moderately skilled: Has performed job independently during training program; limited additional training may be required
4–Skilled: Can perform job independently with no additional training

TASK Complete an accident report. **C593**

Time off__________

Time on__________

Total time__________

1. Use your school's accident report form to complete this task.
   a. Sam, another student, cut himself while helping you service a customer's car. Can you help Sam with the paperwork?

Date	Today
Time	4:45 PM
Name	Sam Smith
Location	Work Bay 1B
Description	A cut on the left thumb
Treatment	Applied first aid. Cleaned the wound and used a Band-Aid to cover the wound. No stitches were required.
Insurance Claim	No

2. Have your supervisor/instructor verify satisfactory completion of this procedure, any observations found, and any necessary action/s recommended.

Performance Rating **CDX Tasksheet Number: C593**

☐ 0 ☐ 1 ☐ 2 ☐ 3 ☐ 4

Supervisor/instructor signature ____________________________ Date__________

Total time________ Time on________ Time off________

Student/intern information:

Name______________________ Date__________ Class______________

Vehicle used for this activity:

Year__________ Make______________________ Model______________

Odometer______________________ VIN ______________________

Learning Objective/Task	CDX Tasksheet Number	2008 NATEF Priority Level
• Identify vehicle needs and obtain customer agreement.	C513	P-1

Recommended Resource Materials

- CDX Automotive program
- CDX eTextbook
- Technical service bulletins, shop manuals, and any other information applicable to the specific vehicle or components you are working on
- Class notes

Materials Required

- Customer (may be role player in simulated environment)
- Vehicle for inspection
- Vehicle labor guide and service information data
- Blank repair order

Some Safety Issues to Consider

- Make sure you understand and observe all legislative and personal safety procedures when carrying out practical assignments. If you are unsure of what these are, ask your supervisor/instructor.

Performance Standard

0–No exposure: No information or practice provided during the program; complete training required
1–Exposure only: General information provided with no practice time; close supervision needed; additional training required
2–Limited practice: Has practiced job during training program; additional training required to develop skill
3–Moderately skilled: Has performed job independently during training program; limited additional training may be required
4–Skilled: Can perform job independently with no additional training

Total time________ Time on________ Time off________

▸ TASK Identify vehicle needs and obtain customer agreement. C513

1. Carry out the following tasks involving personal interaction with a customer while being observed by your supervisor/instructor.
   a. Making appropriate notes, obtain enough relevant information from the customer to enable you to understand the customer's perception of his or her vehicle's condition (customer concern/complaint). Note your findings here:

   b. Provide the customer with advice and information in relation to the following:
      i. The potential courses of action to rectify any issues

      ii. The implications of any course of action (positive or negative)

iii. The most appropriate vehicle repair and/or service procedures for his or her vehicle

iv. The estimated cost of the service or repair

c. Prior to accepting the vehicle for the repair, obtain and record the customer's agreement to the following on the repair order:
   i. A description of the work to be undertaken
   ii. The company's terms and conditions of acceptance, if any
   iii. The estimated cost
   iv. The estimated time of completion, or contact with the customer for an update
   v. Confirm your customer's understanding of the scope of work to be performed.
   vi. Have the customer sign the repair order.

d. Where additional work on the vehicle is required, make contact with the customer and obtain his or her approval prior to undertaking the additional work. Document the agreed course of action on the repair order.

2. Ask your supervisor/instructor to provide feedback and indicate satisfaction with your performance.
   a. Was the advice and information accurate, current, and relevant? Was it presented clearly in a form and manner that the customer would understand? **Yes/No** (Circle one)
      i. Observations:

   b. Did the student demonstrate positive communication skills and actively encourage the customer to ask questions and seek clarification during the conversations? **Yes/No** (Circle one)
      i. Observations:

3. Have your supervisor/instructor verify satisfactory completion of this procedure, any observations found, and any necessary action/s recommended.

Performance Rating

CDX Tasksheet Number: C513

☐	☐	☐	☐	☐
0	1	2	3	4

Supervisor/instructor signature ______________________ Date______

Student/intern information:

Name________________________ Date__________ Class________________

Total time________

Time on________

Time off________

Learning Objective/Task	CDX Tasksheet Number	2008 NATEF Priority Level
• Conduct research to find specific information related to the automotive industry.	C594	P-1

Recommended Resource Materials

- CDX General Service
- CDX Automotive program
- CDX eTextbook
- Class notes

Materials Required

- Service information

Some Safety Issues to Consider

- Make sure that you understand and observe all legislative and personal safety procedures when carrying out practical assignments. If you are unsure of what these are, ask your supervisor/instructor.

Performance Standard

0–No exposure: No information or practice provided during the program; complete training required
1–Exposure only: General information provided with no practice time; close supervision needed; additional training required
2–Limited practice: Has practiced job during training program; additional training required to develop skill
3–Moderately skilled: Has performed job independently during training program; limited additional training may be required
4–Skilled: Can perform job independently with no additional training

Total time________

Time on________

Time off________

▸ TASK Conduct research to find specific information related to the automotive industry. C594

1. While servicing a customer's car, your supervisor/instructor notices that the front tires are wearing out and will need to be replaced soon. Your supervisor/instructor asks Sam, another technician, to take note of the tire specifications so he can get information to inform your customer about the cost and availability. Here is a series of numbers and letters displayed on the side of the tires. Can you help Sam interpret these codes?
   P 205 / 65 R 15 89 H

Tire Code	Explanation
P	
205	
65	
R	
15	
89	
H	

2. Have your supervisor/instructor verify satisfactory completion of this procedure, any observations found, and any necessary action/s recommended.

Performance Rating

CDX Tasksheet Number: C594

☐	☐	☐	☐	☐
0	1	2	3	4

Supervisor/instructor signature________________________ Date__________

Student/intern information:

Name__________ Date__________ Class__________

Vehicle used for this activity:

Year__________ Make__________ Model__________

Odometer__________ VIN __________

Learning Objective/Task	CDX Tasksheet Number	2008 NATEF Priority Level
• Complete a repair order.	C590	P-1

Time off__________

Time on__________

Total time__________

Recommended Resource Materials

- CDX Automotive program
- CDX eTextbook
- Class notes

Materials Required

- Blank repair order
- Vehicle labor guide and service information data

Some Safety Issues to Consider

- Make sure you understand and observe all legislative and personal safety procedures when carrying out practical assignments. If you are unsure of what these are, ask your supervisor/instructor.

Performance Standard

0–No exposure: No information or practice provided during the program; complete training required
1–Exposure only: General information provided with no practice time; close supervision needed; additional training required
2–Limited practice: Has practiced job during training program; additional training required to develop skill
3–Moderately skilled: Has performed job independently during training program; limited additional training may be required
4–Skilled: Can perform job independently with no additional training

▸ TASK Complete a repair order. **C590**

Time off__________

Time on__________

Total time__________

1. Use your company's repair order to complete this task.
   Fred Smith brings in a 2008 Hyundai Santa Fe AWD, with a 3.3 L engine, automatic transmission, 72,426 miles on the odometer, silver paint, VIN 5NMSH73E28H192794. It needs some work before going on a 3000-mile trip. He would like an estimate of repairs needed and has agreed to let your technician inspect the vehicle while you write up the repair order. He gives you the following information.
   a. Home address: 1234 NE Main Street, Anytown, CA 13579
   b. Cell phone: (111) 222-1234
   c. Work phone: (111) 333-4567
   d. Vehicle license number: CDX - 111
2. The customer listed the following concern/complaints:
   a. Small oil leak from under the engine
   b. Small coolant leak from under the engine
   c. Squealing noise coming from the front brakes
3. The technician found the following conditions:
   a. Both valve covers have leaking gaskets.
   b. The water pump is leaking from the shaft and the bearing is worn.
   c. The front brake pads are worn down to the wear indicators, the rotors are a bit under specifications, and the calipers are starting to seep brake fluid past the caliper piston seal and need to be replaced.
4. Complete the repair order as if all tasks were completed including parts, their cost, and labor.
5. Have your supervisor/instructor verify satisfactory completion of this procedure, any observations found, and any necessary action/s recommended.

Performance Rating

CDX Tasksheet Number: C590

☐ 0 ☐ 1 ☐ 2 ☐ 3 ☐ 4

Supervisor/instructor signature ______________________ Date__________

Appendix: CDX/NATEF Correlation Guide for ASE O: Foundation Studies

2008 NATEF Priority*	2005 NATEF Priority*	Tasksheet Title	2008 CDX Tasksheet Number	2005 CDX Tasksheet Number	Page(s)
P-1	P-1	Identify general shop rules and procedures.	C451	C451	1-2
P-1	P-1	Utilize safe procedures for handling tools and equipment.	C452	C452	9-12
P-1	P-1	Identify and use proper placement of floor jacks and jack stands.	C453	C453	13-15
P-1	P-1	Identify and use proper procedures for safe lift operation.	C454	C454	13, 15-16
P-1	P-1	Identify marked safety areas.	C456	C456	1-3
P-1	P-1	Identify the location and use of fire blankets.	C457	C457	1, 3
P-1	P-1	Identify the types and location of fire extinguishers; demonstrate knowledge of the procedures for using fire extinguishers.	C458	C458	1, 3-4
P-1	P-1	Identify the location and use of eye wash stations.	C459	C459	1, 4
P-1	P-1	Identify the location of posted evacuation routes.	C460	C460	1, 4-5
P-1	P-1	Comply with the safe use of safety glasses, gloves, and shoes during lab/shop activities.	C461	C461	7-8
P-1	P-1	Identify and wear appropriate clothing for lab/shop activities.	C462	C462	7-8
P-1	P-1	Comply with appropriate hairstyles for lab/shop activities.	C463	C463	7-8
P-1	P-1	Demonstrate knowledge of safety aspects of supplemental restraint systems (SRS) and antilock braking systems (ABS).	C464	C464	21-22
P-1	P-1	Locate and demonstrate knowledge of material safety data sheets (MSDS).	C465	C465	1, 5
P-1	P-1	Identify tools and their usage in automotive applications.	C466	C466	9-12
P-1	P-1	Identify standard and metric designation.	C467	C467	9-12
P-1	P-1	Demonstrate safe handling and use of appropriate tools.	C468	C468	9-12
P-1	P-1	Demonstrate proper cleaning, storage, and maintenance of tools and equipment.	C469	C469	9-12
P-1	P-1	Proper vehicle identification information: Define the use and purpose of the VIN, engine numbers and data code, locate VIN, apply knowledge of VIN information.	C471	C471	17-18

* These priorities cover assumed knowledge under the 2008 and 2005 NATEF certifications. These tasks are not listed in the required tasksheet, but we recommend inclusion of these tasks in your program.

2008 NATEF Priority	2005 NATEF Priority	Tasksheet Title	2008 CDX Tasksheet Number	2005 CDX Tasksheet Number	Page(s)
P-1	P-1	Identify information needed and the service requested on a repair order.	C472	C472	17-18
P-1	P-1	Identify purpose and demonstrate proper use of fender covers and mats.	C473	C473	13-14
P-1	P-1	Demonstrate use of 3 Cs (concern, cause and correction).	C474	C474	17, 19
P-1	P-1	Review vehicle service history.	C475	C475	17, 19-20
P-1	P-1	Ensure vehicle is prepared to return to customer per school/ company policy (floor mats, steering wheel cover, etc).	C476	C476	13, 16
P-1	P-1	Identify tools and their usage in automotive applications.	C504	C504	9-12
P-1	P-1	Recognize different vehicle types and body components.	C505	C505	23-24
P-1	P-1	Identify vehicle needs and obtain customer agreement.	C513	C513	26-27
P-1	P-1	Complete a job sheet.	C590	N/A	29-30
P-1	P-1	Complete an accident report.	C593	N/A	25
P-1	P-1	Conduct research to find specific information related to the automotive industry.	C594	N/A	28